Corporate Manager's Security Handbook

ANTHONY R. WILLIAMS

AuthorHouse™
1663 Liberty Drive
Bloomington, IN 47403
www.authorhouse.com
Phone: 1-800-839-8640

© 2012 by Anthony R. Williams. All rights reserved.

No part of this book may be reproduced, stored in a retrieval system, or transmitted by any means without the written permission of the author.

Published by AuthorHouse   05/25/2012

ISBN: 978-1-4685-8250-5 (sc)
ISBN: 978-1-4685-8251-2 (e)

Cover Design:
Sabine Solisch
Original form
"Businessman Office" © pab_map - Fotolia.com

Any people depicted in stock imagery provided by Thinkstock are models, and such images are being used for illustrative purposes only.
Certain stock imagery © Thinkstock.

This book is printed on acid-free paper.

Because of the dynamic nature of the Internet, any web addresses or links contained in this book may have changed since publication and may no longer be valid. The views expressed in this work are solely those of the author and do not necessarily reflect the views of the publisher, and the publisher hereby disclaims any responsibility for them.

Endorsements

The "Corporate Manager's Security Handbook" by Mr. Anthony R. Williams is a must read for new managers and a re-enforcement of basic security essentials for today's business community. Security is the keystone of every business in our present world economy. It is short, informative, and to the point. I also like the "Case In Point" segments, which were drawn from Mr. Williams' own experiences, making them more meaningful.

Wayne C. Heinold Sr.
Command Sergeant Major (Retired), US Army Criminal Investigation Command
Asst. Chief of Police/Security Specialist/Cyber Security Practitioner (Retired)

The Security Manager's Handbook written by Mr. Anthony R. Williams, is a masterpiece of collective experiences, hands on experiences, and "the best practice" of the 21st Century security standards. As a former Physical Security Specialist and Fraud Investigator, I must admit this handbook highlights the methodologies and technologies of the security industry. This is a revolutionary approach to security management and standard operational procedures. In an era of corporate espionage, trade secret disclosures, and larceny, this book provides the fundamental steps to understanding the field of corporate security. As Mr. Williams stated "security is the responsibility of every employee, not just the Security Manager." As it pertains to Security Strategies, it is imperative upon the Operational Manager, to understand and be a stakeholder for its implementation. I agree that any violation of security procedures should result in disciplinary action and be documented, to instill a culture of prevention. Mr. Williams presented a list of "major company activities" that was realistic, understandable, and should be adopted into any security procedures. To lead by example, emphasizes how all levels of management should take the seriousness of the security procedures, and do use their position for nonconformity. Mr. Williams identified locations where "Intelligence Gathering" occurs, to expose all employees to the where withal of changing the subject of their conversations to non-related company talk when not on company property. Mr. Williams speaks to the concept of "operation security," which encompasses personnel, property, and data to be adhered with no exceptions. This book is an uncryptic assortment of security information that will provide education, knowledge, and wisdom at your fingertips.

Dr. A.J. King, Sr., Special Agent, US Army Criminal Investigation Command (Retired)

"Security is not an end in and of itself". Security must be an integrated component of a company's success. The Corporate Security Manager's Handbook illustrates in an understandable form how security, through security awareness, must be incorporated in corporation processes. Anthony R. Williams gives many examples from a practical standpoint, and in this manner turns this handbook into a useful textbook for security specialist and everyone in a company that is responsible for security protocols and processes.

Robert Kilian, CFE
Chairperson of the German Chapter of the
Association of Certified Fraud Examiners (ACFE)
Expert for Risk Management (BDSF)

DEDICATION

To Claud Bardon for his quiet and kind inspiration.

To my son Steven through whose eyes I see hope for future generations.

To Sabine Solisch whose artistic vision enhanced this work.

To Sabine Musso for her editorial support

To Veronika Lopez, my tender critic, best fan, and personal motivator.

PROLOGUE

For many years as I conducted security audits and seminars on security awareness operational managers indicated that they needed some simple document to help them be more proactive in security matters. Other managers wished that they had my lecture in written form to assist them in understanding and implementing security measures in their areas. This is a humble attempt at putting a bit of mentor-ship in writing for ready access to specific issues in security from the operational manager's point of view. The text is kept to a minimum on purpose so that the reader is not lost in rhetoric, but can get to the point of the matter.

COMPANY SECURITY PLAN

In general the mission of company security is to insure the safety (from criminal assault) of all company employees, contractors and visitors, as well as making sure that company property is protected against theft, sabotage and vandalism. This definition sounds rudimental, but is complex and the cornerstone of reducing unnecessary monetary losses to the company, while at the same time protecting the corporation's image. Basic as it might seem, if the company security plan is not met, production and profitability, the very purpose for the company existing, will not be at its fullest potential. Furthermore, all those involved with the company will not feel that their assets are safeguarded properly. Ergo, employee productivity is affected, as required work materials are lost, stolen, vandalized, or sabotaged. Hence reducing morale and the willingness to perform well, or protect company assets through individual initiative. On the other hand, visitors, especially customers and suppliers who are not convinced of the adequacy of the company security, will not be convinced that their products and intellectual property are protected in your company. In fact as a matter of practicality, pride, and customer service, a client should never feel the need to request that you improve your security. You should recognize the need and accommodate it without ever being asked.

All of this underscores the importance of developing a concise, functional, and understandable company security plan. Managers at all levels must understand and adopt the security plan into their departments and business units. Moreover, managers must understand that the security plan is not the sole responsibility of the company security manager, but the responsibility of each individual manager. To accomplish this task managers must coordinate with the company security manager to insure that the policies and procedures in their area of control are supporting the company security plan, and that this is understood by each of their subordinates.

Many are of the opinion that company security plan statements are ornamental in nature, only developed for image and publicity purposes, or to reduce insurance premiums. This is far from true. The purpose of the company security plan statement is to help the company as a whole focus on the need for comprehensive security as a means of insuring that the company accomplishes its overall plan of productivity and profitability. This is not to say that security is the single most important factor in insuring production and profitability. But what it should say is that security is a component of the overall business, and is as important to profitability as finance controlling, logistics management, warehouse management and production management. Those corporations wherein the aforementioned components work in concert with the corporate security plan statement in mind are the hard targets that thieves, embezzlers, saboteurs, and terrorists avoid.

COMPANY SECURITY STRATEGY

The company security strategy is the planned security measures implemented to insure that the company security plan is accomplished. That is to say it is all the measures, security force, perimeter control, access control, circulation control, alarm systems, background screening, identification badge procedures, etc, implemented in concert to minimize the security risk to the company.

Of course various internal and external factors are taken into consideration when developing the company security strategy. Such things as weather, geographical location, local culture, labor unions, local laws and regulations, and last but not least, budget; all have an affect on the development of a security strategy. Most important however is that the security strategy be based on the stated company security plan.

Most strategists, be it sports, business, or even military understand the concept that any given strategy is only as good as its weakest facet. Furthermore, that everyone in the organization must understand the strategy as it relates to their individual function, interrelated units, and the proverbial big picture. Managers must therefore understand the company's overall security strategy and how their department or business unit fits into the strategy. Could security breaches in his or her department lead to conditions that would affect production in another department? Could a lack of security in his area result in weaker or no security in another area. Regardless of whether the business units are geographical neighbors or not. A good example of this would be that if a critical area in one manager's department were sabotaged production would stop in several other areas of the company. Another example could be that if raw materials are not well protected, then they will not be available for production causing a work stoppage and increased cost.

The security manager must not only implement and enforce the company security strategy, he or she must, through the corporate chain of command and information distribution functions, disseminate the company security strategy to at a minimum operational manager level. In addition, the security manager must coordinate with operational managers during the development of the security strategy. Too often the security plan and security strategy are developed without input from managers. Therefore it is not surprising that in many organizations, private and governmental, managers and employees do not understand the implemented security strategies. Consequently, support from managers and employees of security procedures are often very low. Many employees consider security procedures as undo harassment, as they do not understand the purpose behind the procedure. Most employees and managers do not understand the purpose of identification badges being worn visibly and security force personnel checking identification badges. That this procedure makes it difficult for criminals to penetrate the company premises and commit crimes harmful to the company is lost in the complaints subsequent to the procedure. In some instances the security procedures are counter-productive or hinder the accomplishment of the overall company business goals.

Case In Point:

As an undercover police officer it was often necessary to infiltrate commercial, private, governmental, and military facilities, which rarely posed a problem as at most locations the employees did not take the access control procedures, even where ID badges were used, seriously. On one of the first occasions I was instructed by my supervisor to gain access to an area that was secured by police officers. A guard was posted every ten meters around the perimeter, I protested that it was not possible. My supervisor told me to observe the access control procedures for several hours and once I discerned a procedural weakness, to exploit it. Much to my surprise I was able to observe a security weakness and used it to gain access to the area. I was acting as law enforcement official, but any other person (thief, terrorist, vandal, etc.) could have done the same as I did.

However, should the security manager involve operational managers in the development of security strategies relevant to the manager's sphere of control, both the operational manager and the security manager can gain important insight. No one will know the inter-workings of the manager's workplace as well as the managers themselves. Such areas as those critical to the production process, tools and equipment subject to theft, weaknesses in the present security system are readily apparent to the manager. This information is of great value to the security manager. Conversely, details of common practices to breach security systems, the best methods to protect critical production processes are second nature to most security managers. The outcome of this process should be that the operational manger has a better understanding of the company security strategy and learns how to evaluate his area of control with regard to security. Of course his acceptance of security procedures will be markedly improved, as will his ability to train and supervise his subordinates with a view towards security.

For the aforementioned reasons operational managers must understand the company security strategy. The understanding must be in enough depth that the manager appreciates the security measures implemented and their necessity in the accomplishment of his business goals. This will afford the manager the insight required to accept responsibility for the security of his/her department. Having taken part in assessing the security strategy for his area the manager will be better able to articulate the need for security to his/her subordinates. Furthermore the manager will come to realize that security is one of his inalienable responsibilities. The manager must insure that security procedures are enforced and followed, include security in his/her departmental training program, and report anomalies to the security manager. Managers must also understand that they are responsible for periodic re-evaluation of the security procedures in their areas of responsibility.

SECURITY IN YOUR AREA OF RESPONSIBILITY

As stated earlier, security in a manager's area is his or her direct responsibility. It is not the direct responsibility of the security manager who only enters the operational manager's department periodically. The operational manager is familiar with the daily operations in his department. They know what should be there and should not be present. Managers will know who should have access to their area and when they should be there. Therefore the first person to recognize an anomaly should be the operational manager.

Many managers have often rebuffed that they have enough to do without being overloaded with security issues. Others are of the opinion that if they enforce security procedures they are acting as policemen or guards. Perhaps a more domestic point of view might help doubting managers better understand why they must be involved in depth in security issues. In the traditional family unit, or where persons share living quarters, it is generally understood that everyone living in the house or apartment has a shared responsibility in protecting the home from fire, thief, and vandalism. Everyone is equally responsible to insure that his or her keys to the home are protected and understand that they will not give unauthorized access to the home to third parties. We lock windows and doors and store valuables in the appropriate agreed upon places. It is also generally understood that when property in the home is lost, damaged or stolen, that it affects everyone who lives there. In addition, all members of the home understand that it is not the job of the police to protect their home directly. The police will assist, give advice, or in some communities they may even conduct a courtesy security inspection. Yet the ultimate responsibility of securing the home lies with its occupants. The same concept is true for the work place. Security personnel are there to assist in security matters indirectly. They secure the perimeter, and advise on procedures. Each employee is responsible for implementing and following said procedures, as in the home, when property is lost, stolen, or damaged in the workplace, it affects everyone-not just the security personnel.

SENSITIVE AREAS

Sensitive areas are those areas where should they be damaged, items stolen, or inoperative, the mission of the company or the safety of company personnel will be compromised. This can be split into two categories. The first of which being simply sensitive which means that the risk is minimal, production will not be stopped, and the health and welfare of employees is not at immediate risk. For example, should someone break into an administration office where no sensitive information is stored, the overall company mission is only at minimal or no risk. Production will not be halted, and employee welfare is not an issue.

On the other hand the second category, critical areas, involve those areas where should they be compromised there will be an immediate halt of production or the safety of company personnel is immediately at risk. A good example of this might be if toxic or explosive gases are used for production, should the tanks be damaged production will stop and the health of employees put immediately into danger. This topic deserves a great deal of consideration, as the ramifications are great, while the potential critical areas are often misunderstood.

Case In Point:

Once in a military environment I was in charge of investigating the theft of an armored tank from a motor park. I noted that none of the remaining armored tanks had any locking mechanisms in them to prevent them from being taken. On the other hand a 2000-liter water tank used to transport drinkable water for troops in the field had a locking device. Knowing that armored tanks cost much more than water tanks I asked the officer in charge why the water tank had a locking device and the armored tanks did not. His reply was that the water tank was a critical item as without drinking water for the troops they would not be able to accomplish their combat mission. In fact he further explained that this was clearly stated in regulations. But with one or two tanks missing they could still move out and accomplish their mission. As funds were short in supply he opted to purchase a locking device for the water tank which he considered critical to accomplishing the mission. The moral of the story is that it is not always obvious as to what items or functions are critical. Careful evaluation must be done to determine this.

ENFORCEMENT OF SECURITY PROCEDURES

Enforcement of security policies and procedures is also the operational manager's direct responsibility. The manager must be attuned to proper use of the identification badges. If a subordinate is not following the policy the manager must make the required correction. Often this is nothing more than reminding a subordinate to put their badge on or to safe guard it properly. The manager must also insure that the doors and windows that should be locked are in fact locked, and not left open for matters of convenience. Some typical examples of why doors and windows are left unlocked or open are: Cigarette smoking, moving shipments in and out, or avoiding having to use identification/magnetic card readers to go in and out of the department. Once again in most cases the manager need only make a minor correction.

However, if the behavior continues, the manager must take the appropriate disciplinary action prescribed in company policy. It is imperative that managers make it known to their subordinates that the manager takes security issues seriously and expects his subordinates to follow security procedures. Singularly most important in this area is that the manager must set the right example with regard to security. Good examples will go a very long way in convincing subordinates that following security policies is not just important, but expected of them.

A bit of old fashioned leadership is required here. It is not acceptable to say, "This came from the top", when enforcing security policies. A stand must be taken to support the security policies as if they were your own. In fact they must be your own. Managers must take ownership of security policies. Enforce security policies in the workplace as if you were in your home. You would not accept a member of your household or a guest not following the security procedures you have established at home. So why accept it in your work place? Remember the lectures and/or sanctions you received as an adolescent when you violated the lock the door policy at home.

Case In Point:

Security forces are often frustrated when senior managers expect to be exempt from established security procedures. So when a manager approaches a security checkpoint, not wearing an ID badge, he expects the guard there to recognize him or her. When the guard stops the manager and asks to see the manager's access badge, the manager gets upset and normally utters something like, "Don't you know who I am?"

This action goes far beyond frustrating the guard. It undermines the security policies in general. Other employees nearby will get the impression that the security policies only apply to non-managers, the managers do not approve of the security policies, and that the policies for security are not important. When these attitudes spread through the organization, the overall security posture will suffer. New policies with regard to security become difficult to implement and existing policies and procedures will be ignored. Moreover, the investment in security hardware may be nullified as employees no longer lock doors, safeguard ID badges, activate security alarms, etc.

TRAIN AND INFORM SUBORDINATES ON SECURITY PROCEDURES

Training is a very important part of any profession or task. No one likes to be given a task to perform which he or she has no knowledge, training or understanding of. This situation almost guarantees that a task will not be performed correctly and all the subsequent consequences thereof. Most of us can remember incidents in our private and professional lives where someone of authority told us to do something that we did not know how to do. If you were brave enough to ask how the task was to be done, often you received a response that was less than helpful. Comments such as "You are just trying to get out of work", or "Didn't you learn that in school" to "You have been working here long enough to know how to do this." If the task was completed correctly it was due to luck and not skill, and probably required a great deal of unnecessary mental and physical effort. Security is too important to be left to the whims of chance and good fortune.

It would be impractical to send each and every employee to formal security training. However, security topics should be incorporated into all meetings. This would only require a few minutes or words detailing existing security procedures and policies. Presented informally, but with great sincerity and conviction. If the delivery of information is half-hearted the employees will feel no need to follow the rules that their manager obviously does not believe in. Then the information is mentally filed under the topic of "Do I know and should I care".

Case In Point:

A state of the art integrated security system is installed in the central headquarters of the security force of a company. The guards responsible for using the system were not involved in selecting the hardware and software installed, nor were they given any training as to how the system should operate. Needless to say there was a great deal of resistance to the system, and the system was not used to its fullest potential, as no one understood the capabilities of the system.

The employees should be instructed at a minimum on how the procedure works, the purpose of the procedure, and whom to contact if they have questions or problems regarding the security procedure. Often demonstrations, if feasible, help a lot in getting the employees to understand the procedure. Furthermore, allowing the employees to practice the procedure goes a long way toward gaining their acceptance and making them proficient at the task. It goes without saying that a task that one understands and performs well will be done freely in the future. It would also help to either distribute the procedure in writing, or to indicate where the written procedure can be found. Remember that a stand must be taken to indicate that you support the procedure. In line with this thought the manager must then insure that he himself follows the procedure at all times.

SET THE EXAMPLE

The old cliché' says, "A picture is worth a thousand words". If this is true then a good example set by a manager is worth a thousand written policies and procedures. Nothing is more discouraging to employees as a supervisor who does not follow company rules and procedures. Therefore it is imperative that managers know all applicable security policies and procedures, and that managers, without exception or excuse, follow the security policies and procedures always. Not when it is convenient, not just when the managers agree with the policies, but always and all policies and procedures.

The temptation is great to ignore certain policies and procedures, but managers expect their subordinates to respect constituted authority and their instructions. Then managers must keep in mind, that corporate security policies and procedures are issued from a constituted authority and are therefore the instructions of the manager's supervisor-not the chief of security. Hence to ignore and disobey security policies is to directly disobey the instructions of their supervisor. This creates a situation in which the manager disobeys the instructions of his supervisor, but expects his or her subordinates to follow their instructions. So the old "Do as I say not as I do" parental guidance that drove us all crazy as teenagers is now in full affect. The value system is distorted and subordinates then feel justified in not following the security policies or procedures. Of course it goes without saying that this could lead to friction between those charged with implementing and enforcing said security procedures. In addition, the security weakness that was at the root of the procedure remains unabated, leaving the company vulnerable to exterior and interior threats. Furthermore, when subordinates face disciplinary action for not having followed security policy, their rationalized defense will be that their direct supervisor does not follow the policy either. It should also be obvious that subordinates, under such conditions, will probably feel no need to follow internal departmental procedures or instructions of their immediate supervisor either. After all, if their supervisor ignores corporate policy, then they have license to ignore the supervisor's policies.

COORDINATE WITH COMPANY SECURITY MANAGER

The company security manager and their staff are the focus point for all matters in the company regarding security. Therefore any and all questions that company managers have with regard to security must be directed to the security manager. So if there is any doubt as to security policy or a potential security risk the security manger or their representative should be sought out. Like all managers, the security manager is very busy, but his job is to protect the personnel, property and interest of the company. Therefore for the security manager there is no such thing as an unimportant question or concern regarding security. In this light it is better to bring a hundred security concerns to the security manager's attention. Even if ninety-nine of these concerns prove to be unwarranted, the one out of one hundred might lead to a cost avoidance in the millions of dollars, or keep a fellow employee (s) from being victimized.

All too often security measures are reactive. That is to say that after an event with negative consequences has occurred security procedures are put into place to prevent the event from happening again. This is all well and good, but damage that could have been avoided has in fact taken place. Therefore in order for security procedures to be proactive, preventing negative events from even happening, company managers must work closely with their security manager. The security manager must be closely involved with the planning of all major projects in the company. Below is a short list of some of the major company activities in which the security manager should be consulted and the interest security has in such matters:

Company Activity	Security Interest
Construction/Renovation	1. Access control of construction workers 2. Security of company property 3. Security of construction can not be locked
Conferences on Company Property	1. Access control of visitors 2. Security of company property
Foreseeable Labor Dispute	1. Access control 2. Additional security personnel 3. Installation of mobile cameras 4. Control of press 5. Personal security/safety of key managers
Internal Departmental Moving Within the Facility	1. Access control in both old and new location 2. Security of sensitive/pilferable company property during the move 3. Change and update of access authorization of employees 4. Access control of mover company personnel
Unscheduled Delivery of Goods and Services	1. Access control of vendor personnel 2. Internal point of contact for goods and services
Major Change in Production/Work Schedule	1. Update access authorization for employees 2. Plan for control of additional deliveries of goods and service. 3. Plan for control of shipping of additional finished product 4. Plan for the security of temporary storage of finished product 5. Plan for the use of more security personnel

It should be obvious from looking at the table above that many of the daily functions in a company generate great many security concerns. Moreover, many of the activities mentioned above, and this is by no means an all-inclusive list, could be situations in which weaknesses in an otherwise good security program will occur. When the security manager has no idea that these events are about to take place, then he or she will not be able to be proactive and adjust security protocols to prevent negative incidents from taking place.

Case in Point:

Lets take a simple example that I have seen in many companies around the world. It does not matter if the company is in the textile industry, food service, or heavy industry. From time to time a department moves to different floor space within the company facility, or may even be moved to floor space outside the company. During the move both employees and company property are moved. Whether the move takes one day or several days inevitably a situation exists wherein company property (Hard property, computers and their components, intellectual property) are left unattended for several hours or even overnight. The property is left in the hallway while employees are not there and the external moving company takes a break. In other instances the old facility, or the new facility contain company property, but neither the old or the new facility are locked, and company property is left there unattended, free for anyone to steal or copy in the case of sensitive documents.

If the security manager is involved in the move he will advise the responsible manager of the precautions to take to effect a move with minimal security risk. Some of the things the security manager will recommend such as first apprising the departmental property in terms of sensitivity and pilferability. Those items prone to theft or sensitive in nature because of copyright, marketing strategy, or research and development, for example, should be locked in a safe place to prevent access by outside movers and persons not authorized to access to the material. The old and new facilities must have a means of being locked, especially if property must be left unattended. Moving personnel should be escorted within the facility and during loading. The inventory of company property must be complete and accurate, and conducted before and after the move. All of these procedures and many others in this situation are often over-looked. This creates unfortunately a golden opportunity for dishonest people to orchestrate negative events.

SECURITY INCIDENT REPORTING

The concise and timely reporting of security incidents is paramount to any functioning security strategy. It enables the security strategy to prevent losses, prevent crime, and gather intelligence that enhances the ability to effectively accomplish the two aforementioned facets of security incident reporting (Loss Prevention and Crime Prevention). Security incidents will vary from suspicious cars parked in front of the office, to the suspected theft of raw materials. What is important is that any and all anomalies be documented properly and reported to the security manager. The format of the report should be kept simple, relating the details of the incident (Who, what, when, where and how) and then stored in a retrievable fashion. Computer databases are ideal for storing security incident report data into an intelligence database. Such reports are the basis of intelligence that can be used to refine the overall security strategy, and prevent losses and crimes.

Of course the question arises, "Which comes first. Loss prevention and crime prevention, or intelligence gathering?" A difficult question, and perhaps there is no clear answer. In my opinion the three co-exist, neither coming distinctly before the other: neither having precedent over the other. That is to say, all three evolve as the situation dictates. During this evolution gathered intelligence helps to prevent losses and crime. On the other hand properly documented losses and crimes generate intelligence that will later help to prevent losses and crimes in the future. In short each facet is as important as the other is and a true symbiotic relationship exists between the three. So one should not try to separate these facets from one another, but learn to establish systems and mechanisms that allow them to interact in a state of controlled flux.

Many managers are of the opinion that the reporting of security relevant incidents in the company is the sole responsibility of the security manager. Many see this facet of management as a police function, a job for which they were not hired. In fact others have referred to this action with the unsavory word of "Whistle-blowing", marking the function to that of an informer or betrayer of trust of friends and colleagues. I strongly recommend that those managers that harbor such thoughts reappraise their commitment and responsibilities in the work place. First of all a manager holds a position wherein leadership and integrity are expected. Secondly, as this is so, it is also implied that managers have a duty to report all incidents that have the potential of harming the company. Of course a manager will feel social responsibility and obligations to family, friends and colleagues; this is only natural. But this should not be in conflict with the manager's loyalty to the company.

Case in Point:

A manager observes a fellow manager on several occasions on the weekend going through file cabinets in an area that has nothing to do with his function. The manager has a personal relationship with the other manager and considers him a friend, but realizes that the activities he has observed are suspicious and could have a negative affect on the company. Of course the manager does not want to be known as an informer, and fears that by reporting the incident he will put his friendship with the other manager at jeopardy, and that others in the company might not like him afterwards. It could be that the actions of the second manager are legitimate, but security should be informed so that the incident can be formally and clearly explained.

LOSS PREVENTION

Loss prevention generally refers to those efforts taken within a business entity to keep its property, raw materials, finished product, equipment, real property, etc., from being lost, damaged or stolen. Some of the basic measures taken to prevent loss are fairly obvious. Physical security methods such as doors, locks, fences, walls and access control helps to protect business assets. Many of these actions are to a great extent the primary responsibility of the company security manager by nature. On the other hand there is a great deal that other managers should be doing to assist in loss prevention. Many of a manager's implied duties, property inventory, property accountability, supervising who has access to the manager's area of responsibility, coordinating with the security manager, and reporting anomalies, all help to deter the loss of company assets under a manager's control.

If one does not exist, a periodic property inventory should be established within a manager's department. Periodic inventories assist in property accountability, while preventing theft and loss at the same time. Too often pieces of real property in an organization, commercial or government achieve invisible status. Invisibility status is achieved when a piece of property is no longer used and placed in a location that is no longer observed or controlled.

The proverbial corner in a workshop, in the back corner of the garage. After a period of time those that work and live in the area no longer really see this object anymore. It stands where it always stands, but no one sees it. Eventually the item is removed from its resting-place, legally or illegally, and no one notices that it is gone until one day when an annual inventory is conducted, or someone needs the item, then the item is discovered missing. Of course no one can recall when he or she last saw this invisible item. In fact often, no one can adequately describe the item, much to the discontentment of law enforcement and security personnel charged with recovering the item. In addition resources are wasted trying to investigate the matter and recover the property. In many instances it is later discovered that someone within the company removed the property legally, but as there was no written accountability and memories are often short, the loaning out of the property was forgotten and a theft reported. If periodic property inventories are done, the phenomena of invisible property can be abated.

Consequently, if unused property must be stored long term on the premises, the manager responsible for the property should coordinate this activity with the security manager. Furthermore, if the manager observes suspicious activities involving such property, the security manager should be informed. This type of information is valuable intelligence that the security manager can use to devise security measures to protect the property at present, or to understand the need for protecting similar property in the future. In addition, this type of intelligence is an important part of developing an overall company security strategy. Therefore an efficient security incident reporting mechanism within a company will greatly help in documenting the aforementioned incidents regarding property accountability so that useful security intelligence can be extrapolated.

CRIME PREVENTION

Many people confuse loss prevention with crime prevention. Crime prevention is, in essence, all those measures taken to prevent crime (Violations of the law) from occurring. A great deal of assets can be lost in a corporation without a crime occurring. For instance, materials being lost or damaged through careless behavior on the part of employees with no specific intent of causing the company harm would not be termed as crimes. Not preparing a production unit properly which leads to the production of goods not being useable (Waste products), could be a costly loss to a company, but no crime has been committed. Improper inventory that leads to goods being stored too long which eventually can no longer be sold as their expiration date has been reached. A costly endeavor, but not criminal in nature.

So when we say crime prevention we mean stopping criminal activity by persons both external and internal to the company. Most crime prevention measures are fairly obvious. Doors and locks, walls and electronic intrusion detection devices all serve the apparent purpose of preventing crime. But other measures are not so obvious. Such things as key-control badge wearing policies, inventories, property accountability, incident reporting, all serve to support crime prevention activities.

Managers play a key role in crime prevention, as it is their responsibility to insure that crime prevention activities are carried out in their area of control. If an incident takes place in their area, with great probability the incident will be brought to the manager's attention, directly or indirectly. The manager must then make sure that the incident is reported via the established method to the appropriate departments in the company. Of course it should go without saying that managers should include crime prevention policies and measures into their internal department training and meetings, and encourage subordinates to be observant with regard to crime prevention and to follow crime prevention policies and measures.

It is not expected that the security layman, be they managers or subordinates, be able to discern the difference between crime prevention and loss prevention. What is important here is that everyone develops a sensibility toward following security procedures, as they are essential to profitability because they help reduce company losses in assets, time and productivity. The importance of incident reporting and all of its facets can therefore never be overstated.

INTELLIGENCE GATHERING

Now we can bring everything together. Security incident reports, crime prevention, loss prevention, coordinating with the security manager. All of the aforementioned facets form the basis for intelligence gathering, which in turn supports crime and loss prevention activities. In order for intelligence gathering to be effective, the reporting mechanism must be simple and easy to use. Long and complicated forms and reporting procedures must be avoided, as well as multiple forms and reporting procedures. A different format or reporting chain for security incidents only discourages employees from reporting incidents. Therefore care should be taken in designing report forms and procedures, and training employees on how to use them. Emphasis must also be placed on the fact that there is no such thing as unimportant incidents reported, and that every report will be evaluated and the information stored. It is also a good idea to give feedback to the reporter of the information in the form of acknowledging the receipt of the information and thanking them for the cooperative effort.

However, the key issue here is the ability of the security strategy to be proactive. When all of the aforementioned security facets are realized, and with the support of managers, the company can be proactive in its security efforts. The old cliché, "An ounce of prevention is worth a pound of cure", will ring true. That is to say that pennies invested in proactive security will yield dollars in cost avoidance. Proper intelligence gathering allows the security manager to apply his security resources in a manner that will efficiently protect company assets, while keeping the interference to the company's daily business to a minimum.

Case in Point:

A young engineer was browsing at a local flee market and notices factory packed products from his company, apparently new, and being sold at an unbelievably cheap price. The engineer purchased one of the products, and then ran test on it and determined that the product was a forgery. Unfortunately he did not notify anyone in the company and the forgery continued.

INFORMATION SECURITY

Information security is still another misunderstood term and often a neglected facet of security. Nowadays most people think of information security as an IT related issue. However, the term information security pre-dates IT. A more inclusive definition of information security might be the mechanisms in place to protect a corporation's sensitive information. This should include virtual information (IT), hard documents, telecommunications, and verbal communications. All of these issues are the responsibility of each employee as well as managers. However, as mentioned earlier, the manager has the responsibility to insure that information security is observed in his area.

IT security is an important facet of overall security, but does not always get the attention and support required from managers. Like many other security policies, employees tend to ignore IT security as unnecessary harassment. Some actually believe that IT support bars employees from having access to certain software or system functions out of laziness or a misguided sense of duty. IT systems in today's business world form the core of its information management. The information in business IT systems is not limited to intellectual property, client account information, or memoranda written within the company management. In many companies production processes are managed by IT systems. All of this information is critical to not just productivity, they are important in maintaining a competitive edge in the industry in which a company competes. Therefore these systems can not be over protected. Especially in view of the ability of hackers to circumvent IT security systems.

The manager must observe IT security protocols, and enforce them in his department. If a member of the manager's department requires increased access rights to IT systems this must be coordinated with the IT security manager. Furthermore the manager must exercise care in selecting who in his department should have which access rights and to which parts of the system. In addition, it is a management responsibility to coordinate the revocation of access rights to a member of his department who no longer needs them, or for a team member that has left his department or the company. IT professionals, especially in large companies, could not be expected to keep up with personnel changes and their relevance to IT access rights. Again, anomalies that occur in IT systems (Spam, chain letters, scams, slowed system reaction times) should be reported as a security incident to both the IT security manager and the security manager (Physical Security). The IT anomaly could be the result of both an IT intrusion or intelligence gathering by criminals for a physical intrusion or other fraud crime (Identity Theft, Nigerian Scam, Bank Fraud, Credit Card Fraud, etc.). Then the appropriate actions including storing the anomaly information as intelligence for use later can be effected accordingly.

Contrary to popular belief hard copy written documents do still exist and are equally as important as their highly technical alleged successor, electronic documents. Therefore they must be protected as well. Remember the clean desk policy? It is intended to prevent the compromise of written/printed documents. A document containing sensitive information that is left open or unprotected can cause as much damage as leaving your computer terminal open and leaving the room. History and Hollywood films are full of stories about spies stealing, photographing, or copying documents under the cover of darkness. The infusion of computers into the workplace has not changed this picture in any way. On the contrary, it has become easier for would be spies (Industrial or otherwise) to access hard copy documents as many people have forgotten that the sensitive document on their computer screen is just as sensitive when they print it out. Managers have the laborious task of insisting that their subordinates protect sensitive documents. If a company document classification policy exist, managers must enforce it. There are numerous tried and true methods to protect sensitive documents; most of them are based on common sense. Such things as:

- Upholding the clean desk policy.
- Using coversheets or water-markings that indicates the level of sensitivity of the document.
- Using a sensitive document register for the receipt of documents taken out of the filing area.
- Restricting who may take a document off the company premises.
- Not reading sensitive documents in public places.
- Not leaving documents unattended while travelling.
- This list is of course not exhaustive, but should provide food for thought with regard to information security as it relates to hard copy documents.

Case In Point:

My plane landed in a major European city. We had to disembark the plane and board a bus to be taken into the terminal. Once we arrived at the terminal everyone hurried off the bus and walked toward the terminal. I noticed a multi-paged document lying on the seat in front of mine, where an elderly gentleman had been seated. I glanced at the document, which was in the English language, and the title of the document indicated that it was a political plan of action of that country's medical association. I stopped the elderly gentleman just inside the terminal and asked him if he had forgotten the document. The gentleman was markedly dismayed. Mutual introductions disclosed the gentleman was on the board of directors of the medical association. The gentleman, pursuant to my polite admonishment, agreed that the document was sensitive, and promised to be more careful in the future.

Probably one of the most successful information security slogans in history in the United States was the U.S. Navy's WWII slogan of "Loose Lips Sink Ships". In spite of hard copy documents and electronic medium, still the form of communication most easily compromised is the spoken word. It does not matter if the word is spoken on a normal telephone, on a cell phone, or between individuals in person. At times the spoken word compromised is more dangerous as there is rarely an audit trail for what one says, and once the word is out so to speak, it is impossible to retrieve it. More over, many people do not realize what they are saying, when they are saying it, and who can hear what they are saying. Therefore those with knowledge sensitive to their company's success, should be aware that "Loose Lips Still Sink Ships". Managers should take special care to self-sensor their verbal communications, be it on the phone, in the office, at home, in transit, or in a meeting.

Consequently, managers must help their subordinates to understand that they too must safeguard against inadvertently discussing sensitive company information in inappropriate situations and locations. Lawful and unlawful gatherers of information have long enjoyed unencumbered information mining from unsuspecting people talking openly about things they should not have been discussing in public.

Below is an unexhausted list of areas and situations where I have often observed sensitive information being discussed inappropriately in centers of public transportation:

- Bus, Train and Air Terminals
- Aboard Public transportation:
  Buses, Trains, Airplanes, Taxis, Shuttle Buses (Rental Car, airport shuttles)
- Public Places:
  Restaurants, sporting events, department stores, hotel lobbies, dinner parties, super markets, doctors waiting rooms, waiting areas in general, movie theater lobbies, private homes.
- At the office/company:
  Lounge, coffee break areas, cafeteria, and meetings, at workstations, waiting areas, and in the smoker's zone.

Case in Point:

While waiting for my plane at the gate in an airport in the United States, I and everyone else at the gate, were disturbed by a middle-aged gentleman in a business suit speaking very loud into his cellular phone about business matters. He repeatedly stated the name of his company and their would be client's company name, as well as the prices he planned to quote in a cost proposal. Aside from disturbing the general peace, the information that he was shouting out could have been very beneficial to his company's competitors, the press or anyone having a desire to harm his company.

Managers and employees must remember that company information is entrusted to them and intended to be used for the benefit of the company, not as entertainment or as chitchat filler. Care must be taken, even at company meetings, to insure that sensitive information is not discussed when persons not authorized to receive said information are present. It must be clear before a meeting, what information is to be discussed, who will be present, and what level of information each person is authorized to receive and their legitimate need of the information to perform their specific task for the company.

As a self-test, the next time that you are with colleagues at a social function, lunch, dinner parties, trade fairs, in transit, listen closely to the conversations. Try to discern how many times information is discussed that were not intended for public dissemination. Next survey the location, and count how many people was present that should not have had access to the information discussed. In actuality it is natural for humans to talk to one another. The spoken word is one of the blessings of humanity. Then again there are so many interesting topics and events happening in our lives that we should be able to converse about them and avoid discussing business matters. It is a good idea when frequenting with business associates, to set a rule that you will not "Talk Shop". It requires a great deal of self discipline at first, but one can hold conversations and not talk about what is being done at work.

Case In Point:

While lounging about in a bar near a production facility where I was scheduled to conduct security training, I listened to the conversations of employees from the production facility as many vented their frustrations of instructions or actions of their supervisors that they felt were wrong. Others discussed overtime that was necessary to meet production schedules or detailed what they believed to be company weaknesses. All of this information was strategic in nature and very useful to the media, competitors, terrorist, and thieves.

TRANSIT SECURITY

In the global economy business travel is a necessity and may consist of local, regional, national or international travel. Aside from the strain of normal travel, business travel presents some unique problems, many of which are rarely thought of, and therefore rarely prepared for. The major problem areas associated with business travel are industrial espionage, political environment in foreign countries, and crime in general. Falling victim to any of the aforementioned areas could compromise your firm, cause a loss of business, the loss of personal and or company property, and last but certainly not least personal injury. Therefore security must be a part of planning business travel.

First of all for any planned travel outside of the immediate area where you work, you should ascertain the crime and security trends for that area. The web site of police organizations in the area that you plan to visit is a good place to start. The police departments in most major cities have homepages that list criminal statistics, often including crime maps. This information is useful for booking hotels, picking restaurants, planning travel routes within the area, and selecting rental cars.

Where international travel is concerned most countries provide information regarding crime, tourism, terrorism, local customs, business and banking practices on the homepage of their Consulate General. It is in your own best interest to review such material while planning business and even private trips abroad. Never be afraid to ask the organization that you plan to visit about the aforementioned topics prior to travelling. This will insure that your trip is not marred with unpleasant experiences that prevent you from conducting business productively.

Case In Point:

While on business in the middle east one of my associates while at dinner forgot the health warning that we had received from our host and ate the raw vegetables, that were meant as decoration, from his plate. He became very ill and was not able to work for several days. His experience was not only painful for him; it caused extra work for the rest of the team.

The selection of hotel accommodations is equally important, especially in large cities and abroad. Comfort and cleanliness are essential to conducting effective business, and most people take this into consideration when they book hotels. However, security is equally important. This begins with the location of the hotel, for if the hotel is located in a high crime area, your personal safety could be threatened and you are at risk for theft, assault and robbery.

If your business travel takes you to a city or area with a high crime rate, it is suggested that you select a hotel that operates its own access control and security. But no matter what hotel you are staying in there are several areas one must take care in. Never leave your notebook; cell phone, PDA, or other electronic business device unattended. They are often the target of illegal information brokers and thieves. Thieves are interested in quick money, but information brokers are interested in the data stored on the aforementioned devices, which more often are much more valuable than the device itself. Furthermore, information brokers and industrial spies might manipulate unattended electronic devices in a manner that makes them gather and/or transmit information to unauthorized persons. Inserting "Trojan Horses" in notebooks, or listening devices in cell phones is neither uncommon nor difficult. The use of the safe in the hotel room is also ill advised. These safes will keep honest people honest, but they will not stop an experienced thief or industrial spy. Furthermore, often hotel personnel have access to the safes in hotel rooms.

If it is absolutely necessary to have these devices on business travel, then they should be with you at all times. This also means at meals and while travelling in the area, your devices must be with you, under your direct control. Leaving a personal computer in cars, to include the trunk of a car has made many personal computers easy prey for thieves. Strong consideration should be made before travelling with electronic devices, and the data stored on them must be documented at your home base. In this manner, if the devices are lost or stolen, a record of their contents is available and the damage to the company because of their compromise can be properly assessed. Ideally, the hard drive of company owned computer notebooks should be encrypted and protected with password and token.

In fact it might be wiser when visiting components of one's own company to access your business data from the company server using a workstation at the location you are visiting. A second remedy would be to travel with USB data sticks containing the information you need for work. This would make travel in the area easier, but great care would have to be exercised to prevent its loss, especially if the USB stick is not encrypted. The USB sticks should be kept separate from the notebook, that is to say, not stored in one of the pockets of the notebook case in which the notebook is carried. Carry the USB sticks preferably directly on your person. With regard to the notebook, if at all possible while in transit, carry your notebook in a case that does not look like a notebook case. Many thieves purposely look for notebook cases when selecting their victims. A pilot's case is a good example of a container that can be used to carry a notebook that does not broadcast the fact that you are carrying a notebook. Other containers such as backpacks, briefcases, etc. are also suitable camouflage for notebooks.

SECURITY AT CLIENT/ CUSTOMER PREMISES

It goes without saying that while visiting the premises of clients, customers and other business partners, courtesy demands that you respectfully follow the security procedures of your host, so long as these security procedures do not conflict with your company security procedures. However, caution must be exercised with regard to information security to protect your company's sensitive information.

Below is a list of "**Do Not**" while visiting the aforementioned premises:

- **Do not** enter the host's IT networks using your company laptop computer.
- **Do not** use USB storage devices provided by the host on your company laptop unless your company has a procedure for doing so.
- **Do not** leave your company laptop and other electronic devices unattended.
- **Do not** leave sensitive company documents unattained.
- **Do not** take sensitive company documents with you if it is avoidable.

- **Do not** discuss sensitive company information that the host is not authorized to have.

Following sound security practices while visiting clients and customers might irritate some people when they can not get additional information from you, even when they know that they should not have the information. However, most people that you conduct business with will appreciate and understand your precautions, as they are good indicators that the information that they provided your company (Intellectual Property, etc.) is well protected. This builds trust, and supports good business relationships.

COORDINATING SECURITY OUTSIDE OF THE COMPANY

No company or business lives in a vacuum, they are surrounded by other businesses, residents, etc, all with different goals and needs. Hence the environment in which a business is physically located can not be ignored or underestimated. Many companies spend a great deal of time and money in the location selection process, in hopes of positioning the company in an environment conducive to profit. However, all too often the security aspects of a location are not taken seriously into consideration. Crime statistics for the area, police operations and the core business activities of direct and indirect neighbors, are as equally important as infrastructure and parking space for customers.

Crime statistics for the area in which the company is located bears on security and the image of the company. It goes without saying that customers and employees will not feel well conducting business in a high crime area. In addition, insurance premiums can be affected with regard to liability, theft and vandalism. It is also obvious that the cost of security is directly related to the crime rate in the area, as the crime rate and types of crime in the area form the basis for perimeter security, access control and intrusion alarm devices. Therefore security managers must insure that they have access to current and reliable criminal statistics. Current crime statistics can also be helpful in determining security guard manpower and hours of operation.

Obtaining reliable crime statistics is not always simple in some countries, but can be easier if there is effective coordination with area law enforcement. A positive working relationship with the local police will also insure that whenever the company or its employees/visitors are victimized by criminals, on or near the company premises, the police provide prompt and efficient service. The police should also be coordinated regarding to construction projects that might affect vehicle traffic in the vicinity of your company, regardless as to whether the project originates from the city, a neighbor or your company. This reduces traffic tie-ups while increasing driver safety. Furthermore, in times of social, economic, and labor disputes, coordination with the police can lessen the negative effects on your company, such as demonstrations, vandalism, assaults, and disruption of production.

SMALL BUSINESS AND SECURITY ADDENDUM

Naturally all of the facets laid forth in this text will apply in part or full to a small business, no matter how small a business, as these facets are universal. The primary difference in applying this text to small business is implementation. The personnel restraints of small companies may not have the budget to support a full time chief of security who is devoted solely to security matters. In other cases smaller companies are tenants in larger industrial or business parks, with the primary responsibility for external security belonging to the facility owner or the owners designated organization. Constellations also exist where small firms are structurally independent, but share borders with other companies. Whenever one of the aforementioned conditions exists, it is strongly recommended that someone within the company be given full authority and responsibility for security. Such a position is challenging, but necessary for the well being of the company. It goes without saying that the security duties will be in addition to normal operational duties, but it is important to have one person responsible for security, even in a part time status. Key problems for small businesses with regard to security are cost, developing an understanding of the need for security, and coordination with neighboring companies and local authorities. One person dedicated to the company security has a better chance of monitoring and resolving these problems.

COST

Contrary to popular belief, security in small businesses need not be expensive. Using the same principles mentioned in this text is actually easier because small businesses normally have staffs that are relatively small, accessible, and easier to control than that of large corporation. Furthermore, the facilities of small businesses tend to be much more compact than that of corporations. Consequently, the cost of security systems is proportionally reduced. Many of the measures required to increase security in small businesses are minor measures governed by common sense.

On the other hand, small businesses by their nature can not afford the financial cost associated with theft, vandalism, sabotage and litigation regarding intellectual property and patents. In this light security is more critical to small businesses.

Case in Point:

A small production and retail beverage operation with less than ten employees had experienced a couple of break-ins in a short period of time. The police authorities quickly identified and arrested the juvenile delinquents that were responsible. The owner sought the author's assistance, which only required two hours of time and some down to earth discussion. The owner then purchased and installed a very basic security system from an electronic distributor to protect his facility during non—business hours. Customer access procedures were evaluated and changed, product displays and products in front of the facility were also changed in a way that discouraged intruders by eliminating areas that an intruder could have hidden while breaking into the facility. Additional lighting was also installed outside of the building and the cash register was moved to a better location that would discourage a visitor who entered the retail area from taking money out of the register. None of these improvements to security were expensive, and the owner did most of the physical labor, and the break-ins stopped.

SECURITY AWARENESS

Awareness of security is imperative for small businesses, as losses in small businesses are more painful because their resources are much more limited than that of corporations. When a small business has losses, all members of the business feel the effects much quicker than in large businesses, and survival becomes more difficult. Security guards are rarely affordable, and if customers, suppliers and business partners, employees feel unsafe, they will avoid doing business there. However, small businesses have the advantage of being small. That is to say, that direct communication to all members of the business is usually very easy and more times than none, informal. Hence the need for security is easy to disseminate, as well as the need for everyone to do their part to protect their common interest, their jobs and their safety.

COORDINATION

Coordination of security efforts is a critical issue in small business. Coordination can reduce the cost of security and improve the relationships with neighboring businesses, local government and policing agencies operating in the area.

The old cliché, "A chain is only as strong as its weakest link", can be the key to effective area security for small businesses. Many small businesses are located in industrial/business parks, office buildings, residential and retail zone areas. In all of these locations the security effectiveness is dependent in part on one's neighbors. One business may have very strong security for its premises, but if the neighbor has weak security it affects your business as intruders may access your premises through your neighbor, after having broken into the neighbor's facilities. It is similar to a residential apartment building that uses a buzzer system to allow visitors to enter the building. When one resident ignores the rules and buzzes everyone in without checking to see who it is, then would be thieves enter and could break into the apartments of residents who follow the rules.

Therefore it is important to coordinate your security efforts with all your neighbors. If a consensus can not be reached, then you must take the required measures to increase the security in the area bordering the uncooperative neighbor. On the other hand, if an agreement can be reached with neighbors, resources can be pooled to purchase intrusion detection systems and guard services for full or part time use. On occasion industrial parks provide security systems and/or guard service as part of lease agreements. Again good coordination with neighbors and/or the owner of the complex can help greatly in insuring that the security measures are effective as well as economical. However, if the complex owner does not offer security services, then requests by all the tenants for security services will not be ignored.

Communities are driven by commerce. Businesses provide desired goods and services to a community and pay tax revenue. Most people want to live in an area where goods and services that enhance the quality of life are readily available. It is the responsibility of local government to make sure that the relationships between businesses and the community are healthy, because when the relationships are healthy, then communities thrive and grow, and citizens are happy. Small businesses must learn to coordinate their security needs with local government. This is not to say that local government is to pay for security services, no indeed. Security does not start and end on the premises of a small or large business, but in the environment in which they are located. James Q. Wilson and George L. Kelling's "Broken Window Theory" explains that communities in decay tend to attract criminal activity.[1] Things like deserted run down buildings, massive graffiti, broken street lamps, and dirty streets are a sign to criminals that the community is not unified and will probably not resist criminal activity. Local government is the focal point for correcting these issues and for their own benefit. Because if the area deteriorates, then both businesses and residents will move out of the area, the community will cease to thrive, property values will go down, and tax revenues

---

[1] James Q. Wilson and George L. Kelling, The Atlantic, Broken Window, 1982

will decrease. Of course local government officials will not get re-elected under these conditions.

Law enforcement agencies are generally under the influence of local government, even in countries with multi-leveled police structures like the United States. However it is more effective to coordinate directly with police agencies whenever possible. In larger metropolitan areas police agencies may have a division dedicated to assisting local businesses, and may even offer free security assessments. Close coordination with the police will assist the police in providing meaningful routine services to you, such as periodic patrol, rapid response when problems occur, and reporting general problems such as malfunctioning street lamps, garbage pick up, traffic patterns, to local government for resolution. Furthermore, the police will have a great deal of information regarding crime trends that will be useful to you in developing your own security strategy. Local police can also give recommendations with regard to the reliability of security guard services. Lastly, local police share interest in insuring the quality of life in their area of responsibility, and preventing crime that can lead to the decay of a community. Therefore the welfare of businesses in their area is very important to the police.

USE OF SECURITY CONSULTANTS ADDENDUM

Security consultants provide objective opinions regarding the security posture of an organization. Furthermore, a security consultant can assist you in correcting security deficiencies and developing a security strategy tailored to your company. The question is often asked, "What do I need a security consultant for when I have an internal security organization"? The answer is simple, too often the stress and strain of the normal workday, and the pneumonia of seeing a facility every day distorts one's vision and impairs objectivity. Security consultants are immune to this pneumonia and are in a better position to see things as they are. This is a very important aspect as it is natural for our perceptions to cause us to stop taking notice to certain security issues.

Case in Point

While conducting a security audit of a production sight the perimeter is surveyed by walking along the interior and exterior of the location. The chief of security was of course present. While examining the rear perimeter which consisted of a tall chain link wire fence mounted on a brick wall that was roughly knee high, the security consultant noticed that on one section of the brick wall the mortar appeared loose. Physical examination revealed that bricks were not secured by mortar or cement, and could be easily removed which afforded a point of entry to the facility large enough for a grown man. This security weakness was compounded by the fact that this portion of the perimeter was at the rear of the production facility, directly on a road that had little traffic after normal business hours and no intrusion detection devices or CVC cameras protected this area. The chief of security stated that he routinely checked this area, but had never noticed that the bricks could be removed.

No one knows everything, as it is not possible to stay abreast of all the changes in any industry or discipline. This is not to say that all consultants will have more knowledge than security managers will. But one can say that the professional security consultants have a better chance of staying abreast to changes. Furthermore, the security consultant's focus on analyzing security in varied industries helps him or her to develop a wealth of knowledge through experience. Consequently, security consultants share their knowledge with their clients. This is very important for small businesses that either have no one directly responsible for security or an inexperienced person in charge of security. Hence the mentoring function of security consultants can prove invaluable, especially to small businesses.

Case in Point

A security audit at a small energy production facility was conducted after the firm had invested a great deal of money in intrusion detection devices and CVC cameras. Unfortunately the equipment installed was grossly inadequate. Cameras and alarms were installed in locations that made them completely ineffective. The cameras were positioned in the wrong locations and in the wrong angles. Moreover, two employees whose primary function was to monitor images on much larger screens that documented the automated production system monitored the images generated by the cameras after hours on small computer screens. Furthermore, the CVC cameras were not augmented with motion detectors or an audio alarm. Consequently, if the persons monitoring the system were not looking directly at the monitor when the CVC camera was capturing an intruder's image, the intrusion would go unnoticed. In the final analysis a great deal of money was spent for devices that did not perform the desired function, and the security of the facility was not enhanced. Unfortunately in such a situation the owners of the facility have a false sense of having adequate security. After all, they have a new and expensive security system.

Needless to say an experienced security consultant would have been able to assist in purchasing the right equipment and having it installed correctly. Often firms that sell security devices offer their advice on what equipment is needed and where the devices should be installed. While many of these companies are honest and operate under good faith principles, the opinion of a disinterested third party, the security consultant, can insure that you actually buy what is needed and not what a company merely wants to sell.

The presence of an outside security consultant performing a security audit is a useful tool in displaying to employees that security is taken seriously in a company. In addition, employees may take the opportunity to address questions about security to the consultant, or express their concerns about security. A wealth of information, not readily known to the company management or security organization, might be revealed to the security consultant.

Case in Point

While security consultants were interviewing employees about their security concerns at a production facility, a young employee mentioned that he/she thought that it was strange that apparently new merchandise containing components that were produced at their facility were being sold far under price at a local flea market. Acting on the information it was discovered that components were being stolen in large scale and sold in the black-market to counterfeiters. This condition of course was causing the client company to lose sales. As no one had ever spoken with the young employee, and the employee had no idea who to speak to about the matter, had the consultants not interviewed the employee, the condition would have gone on indefinitely.

It is not uncommon that employees observe anomalies in the course of their daily business or private activities. It is also not uncommon that employees, not knowing to whom or how to report anomalies, fail to report valuable information to their employers or security personal. There are many reasons for this. More times than none the employee does not realize the significance of their observations. Others understand the gravity of what they have seen, but are not aware of avenues in their organization by which they could report incidents. Lastly, the employee might consider many observations, as a matter of general knowledge that they feel the company already knows about. Therefore it is not surprising that often such information is uncovered when a skilled security consultant interviews employees.

Furthermore, many employees are not aware of the communication channels within their corporation for reporting their concerns regarding security. Conditions such as fellow employees circumventing security systems, a perceived security weakness, or employee suggestions to improve security procedures many times remain unspoken until someone asks the right question. Again, the security consultant is in the position and has the experience required to ascertain when and what questions to ask employees.

EPILOGUE

Security at the operational management level remains a widely misunderstood part of governmental and commercial enterprises. Understanding one's role in security is very difficult at all levels of corporate life. Part of the confusion is attributed to the tendency of many people not to accept first the need for security, and then not accept responsibility for security in general. It is hoped that this text will help operational managers understand the reality of security and how it affects their work environment. Consequently, understanding the reality of security will help the manager work in concert with security professionals to prevent loss of property, insure the safety of their subordinates, and deter those who would intentionally cause harm in the manager's area. Lastly, this text should help the operational managers and owners of small businesses realize that they are not alone in their business environment and that close coordination is required with their neighbors, the police, local politicians, and other divisions within their organization, to keep their businesses safe and profitable.

REFERENCES

- James Q. Wilson and George L. Kelling, The Atlantic, Broken Window, 1982.

Photo Credits

Book Cover

- Businessman Office © pab_map – Fotolia.com
  ID # 11967997

Book

- Zahnräder 2 © arahan – Fotolia.com
  ID # 3017844
- wooden chess game pieces © Bruce Shippee – Fotolia.com
  ID # 2301357
- Wicket © Jerzy Opoka – Fotolia.com
  ID # 33597927
- Wassertropfen1 © fotofuerst – Fotolia.com
  ID # 18450637
- Man Portrait intellectual holding glasses
  © snaptitude – Fotolia.com
  ID # 23399406
- businessman in front of computer © knipsit – Fotolia.com
  ID # 14301551

- moving house van © flashpics – Fotolia.com
  ID # 21261513
- Man in a corridor © Yury Umyvakin – Fotolia.com
  ID # 19353066
- flohmarktstand (9) © fuxart – Fotolia.com
  ID # 853561
- Secret Documents © pmphoto – Fotolia.com
  ID # 4034078
- man listen © mast3r – Fotolia.com
  ID # 25956808
- Businessman talking on mobile phone © Monkey Business – Fotolia.com – ID # 7795160
- Cafe scene © corepics – Fotolia.com
  ID # 12811355
- cucina tipica araba © zonch – Fotolia.com
  ID # 27283198
- Antik © Martina Berg – Fotolia.com
  ID # 4987424
- Gebrochene Wand © Bernd Kröger – Fotolia.com
  ID # 15374188
- Kamera – Überwachung – Sicherheit © Tiberius Gracchus – Fotolia.com – ID # 32322823
- Business – Job Interview © Kzenon – Fotolia.com
  ID # 33340530

www.ingramcontent.com/pod-product-compliance
Lightning Source LLC
Chambersburg PA
CBHW021007180526
45163CB00005B/1923